WHO'S IN CHARGE?
WILDERNESS CHANGE
AND EVOLUTION

WHO'S IN CHARGE? WILDERNESS CHANGE AND EVOLUTION

(a snap-shot of life on earth)

BILL WONDERS

AuthorHouse™ LLC
1663 Liberty Drive
Bloomington, IN 47403
www.authorhouse.com
Phone: 1-800-839-8640

Published by AuthorHouse 02/07/2014

ISBN: 978-1-4918-6074-8 (sc)
ISBN: 978-1-4918-6043-4 (e)

Library of Congress Control Number: 2014902227

This book is printed on acid-free paper.

Table of Contents

Drawing and artwork

The artwork and drawings presented in this book add significant support to the message in text. Truly a picture can say 1,000 words. I am in debt to my son Eric and Emil (Butch) Perunko (Art-4-Ever) for the following drawings:

Eric Wonders

Emil (Butch) Perunko

Who's In Charge?

Wilderness change and evolution

(A snap-shot of life on earth)

Opening remarks:

Ever stand in a forest at dark, harboring a feeling that you were being intensely watched, evaluated, or sized up by an unknown, unseen, predatory wild animal? Ever give thought to human physical and/or mental frailness compared to predatory wild animals? Ever look directly into the eyes of a predatory animal (caged or free) as it stood motionless, intently staring at you, evaluating every move you made? Ever consider having a wild animal as a household pet? Ever have occasion to reflect on how humans as well as all animals fit into earth's long-running evolution? If you answered yes to any of these questions, this book is for you.

Who's In Charge was not written to frighten or drive away outdoor enthusiasts, hikers, campers, hunters, or anyone interested in exploring the North American wilderness. Just the opposite, my personal goal is and has always been to encourage more people to get out, experience, and enjoy the outdoor wonders of this great land. *Who's In Charge* was written to provide thought regarding inherent wilderness dangers, predator encounters, the changing environment, and, most important, a snap-shot of the big, overall, picture of evolutionary history and change on earth.

Venturing into the wilderness today a person needs to be prepared for a changing world, what I refer to as a "new order" in the wild. The environment is simply different and animal behavior changes are clearly visible. The changed behaviors noted in this book may also be a small component of a much bigger picture.

Asking who is responsible or who is in charge is my question. Let me present some facts, tell true stories, and give examples of this new order. Readers can draw their own conclusions.

In Memory

Eric Wallace Wonders

1978-2010

Acknowledgements and appreciation

My personal and heartfelt thanks go out to my family and friends who have supported me and helped make life on earth so rich and rewarding. My two sons, Will and Eric, earned and deserve my greatest love and most sincere appreciation. Will has served as an ongoing inspiration through difficult times and Eric, who we lost all too soon, for his beautiful art work and lasting memories; he will rest forever in my heart.

A special thanks also to Emil (Butch) Perunko (Art-4-Ever) for his excellent art contributions and to Nancy E. Piazza, Writeperson Ltd—Western Reserve Editing for her assistance and support.

Lions or Gazelles?

"Every morning in Africa, a gazelle wakes up.
It knows that it must run faster than the
fastest lion or it will be killed.
Every morning in Africa a lion wakes up.
It knows it must outrun the slowest gazelle or it will starve to death.
It doesn't matter whether you are a lion or a gazelle,
When the sun comes up,
You'd better be running."

Business management publication

Author unknown

Introduction

(Why write this book)

"Life is our dictionary . . . .
Colleges and books only copy the language
which the field and work yard made."

Ralph Waldo Emerson

A solo exploration adventure into Central Alaska provided an animal encounter that brought to the fore a changed wilderness environment and "being the hunted" into clear, frightening perspective.

The late summer day was absolutely beautiful, blessed by Alaska's tremendous expanse, great fresh air, and unparalleled outdoor sights. I drove my rented car off road into an area I had previously visited—an area that provided spectacular views of the "Great Land" in Central Alaska. The mountain I elected to climb this day presented a rocky incline where my assent ended on a precipitous rock outcropping above the tree line. This spot commanded an excellent view of a huge valley running north and south through the Alaska Range. The climb was not inconsequential, as the grade was steep, plus it was a windy day with some areas offering rather poor footing. It took nearly one hour to reach the rock outcropping I targeted. This special point on earth called to mind the magic of Robert Service's poetry. Every direction on the compass rose provided a scenic view that would fit well on an outdoor calendar.

"It's the great, big, broad land 'way up yonder,
It's the forests where silence has lease;
It's the beauty that thrills me with wonder,
It's the stillness that fills me with peace."

Spell of the Yukon
Robert W. Service

From my vantage point I could see for miles up and down an impressive valley, the river drainages, alluvial fans, endless rock slides, all formed at the base of several mountains linked together in the Alaska mountain range. It was humanly impossible to properly reflect on the beauty of this area or to fully comprehend the titanic plate tectonic forces forming these massive mountain peaks.

Using binoculars, I scanned the mountain sides, top to bottom, absorbed in the beauty of this wilderness, void of other humans. From my vantage point I commanded an exclusive view of this magnificent terrain that included a few caribou, a cow moose with a yearling calf, numerous ground squirrels, and eventually one monster of a brown bear. This bear was in open territory approximately three or four miles distant. He was one of those giants with an attitude. He took his time, walked where he wanted, and did what he wanted, when he wanted. He was a magnificent animal.

The wind continued to be stiff, blowing directly down the valley ahead of the bear. Many outdoor experts (including the late Timothy Treadwell), and a few national park rangers said that wild bears run for cover at the first scent of a human. It was clear that the direction this monster was going would soon bring him directly to where the wind carried my human odor. I kept my binoculars fixed on the bear as he meandered over the tundra, and bingo, he caught my scent just as anticipated. The bear sat back on his haunches, continued to smell the air looking around, soon fixing his stare directly at me. No doubt the bear smelled me then saw me, even if only a dot on the side of a distant mountain. The experts were wrong. This bear did not run for cover; he continued to sniff the air and abruptly changed direction, heading slowly, straight for me. Through my binoculars, I was looking at this bruin head-on, directly into those steel-dark eyes as he moved in my direction. Clearly it was time to depart. The sun was up and I was running.

My car, near the base of this mountain, was parked between the bear and me, yet much closer to me. If the bear didn't charge, I knew with a steady decent, I'd reach the protective safety of the car ahead of the bear. If I provoked or gave this bear incentive to charge, I wouldn't stand a chance. Bears can run over rough terrain at over thirty miles an hour. I on a good day, on flat land, with running shoes and a wind at my back, might zip along at eight or nine miles an hour. No match here, and I was not the animal in charge. Slowly, carefully, I stood and started a careful, determined decent. If I ran, it might provoke the bear into running, and he would certainly intercept me before I reached the base of the mountain and safety of the car. I did manage to get behind a rocky ridge, out of the bear's line of sight, and accelerated my decent, however my escape required me to reappear in open sight of the bear before I reached the cars' protection. As I came back into the bear's sight, I was relieved to see he was not running. But he was still moving directly toward me at his slow deliberate pace, still sniffing the air, determined and focused—focused on me. I remember thinking I wished one of those "bear experts" were in my shoes right now. Their bear/human scent philosophy might change as this bear just got bigger and bigger and closer and closer.

About the time panic started to take hold, I reached a point in my decent where I felt confident I could reach the safety of the car ahead of the bear and ran like a gazelle. After opening the car's door, I stopped and saw that the bear had also stopped, he continued looking directly at me although the distance separating us was now yards not miles. He sat back on his rear haunches, looked around, apparently decided that I was now more trouble than I was worth, and then turned, leisurely walking back in his original direction with an occasional look back at me. I was frozen in place watching this monster move away. At this point I also realized that I was in charge of this confrontation. In charge only because I held on to an escape route (the rented car), absent that the bear was in charge.

No question in my mind, then or now, that big bear smelled me, saw me, held no fear of me, and decided I might make an easy dinner. Nope, no question in my mind.

It was this experience that first got me thinking about a "new order" or evolutionary change in the wilderness and served as the genesis of this book. Was my encounter unique or becoming more common? Why were bear experts so confident in predicting bear behavior when I could see first-hand that their behavior could be anything but predictable?

I had a chance meeting with an interesting Alaska State Biologist in Anchorage a year later who talked with me about what I referred to as a new order in the wild. This biologist worked for the Alaska Department of Fish and Game and had just transferred from an assignment in Yakutat, Alaska. He was a seasoned Alaska hunter and outdoor person who had experienced numerous bear encounters. He agreed that generally wild bears did everything they could to avoid humans and would disappear at the first human sight or smell. That was most of the time, not all. He shared with me a few of his bear encounters, including an experience he had when hunting black tail deer on the Alaska Panhandle. On this fall day he was following a small stream through a thick area of woods and came upon a big brown bear sitting in the shallow water at the side of the stream. Here too the bear was a huge one with an attitude and displayed no intention of departing his spot on the narrow stream. This was an area where bears had been hunted for decades and "generally" avoided humans at all costs. Consequently, the biologist decided to shoot his high-powered rifle at some gravel directly in front of the bear to frighten the bear, get the bear moving, and allow himself to pass. He shot and hit gravel right adjacent to one of the bear's front paws. The rifle rapport was loud, gravel few everywhere from the bullet strike, however the big bear didn't budge. This big boar turned his huge head and looked directly at him as if to say, "Just who the hell are you?" Here too the biologist wisely decided, "Hey, maybe it's time for me to leave." He cautiously, avoiding direct eye contact, slowly backed away and found an alternate (albeit more difficult) route around this creature. Clearly this big wild bear held no fear of humans.

I will submit that there are subtle, yet dramatic evolutionary changes in the North American Wilderness: animal habitat, physical geography, food availability, and operational boundaries. While these changes may undermine the historical balance between man and nature they are simply a small part of overall animal and plant evolution where humans represent a microscopic segment of the universe.

This book is designed to get readers thinking about the new environment, the new dangers faced, the new relationships that have evolved (some with deadly consequences), and the big picture of evolution on earth.

Humans simply do not command the highest rung on the food ladder and wild animals, large and small, have developed a new attitude regarding humans. Humans are not seen as a deterrent to territorial expansion as in the past. Wild animals simply adjust to human interaction and urban sprawl (evolve). Frail humans can and do present a new food source. *Who's In Charge* reflects on these changes or a "new order" in the wild, with thoughts on the big picture of earth's evolution and our microscopic role in the universe.

Chapter One

Evolution and Adaptation

(The way of life on planet Earth)

Life on planet earth has been in a constant state of change, evolution, and adaptation from its very beginning. Our delicate planet was built a few billion years back when a host of cosmic collisions took place at just the right time, in just the right place, and in the right order to nurture life. Once the right biological and environmental conditions were in place, animal life on earth began (evolved) in the sea around 500 million years ago with tiny creatures like the trilobite. The trilobite of the Cambrian period may have been the first truly living creature on planet earth.

Later, in the Silurian and Devonian periods, animals, like acanthostega, evolved. They gradually ventured from sea to land, breathing air and challenging gravity to begin life above water. Soon the flippers were gone, replaced by legs and feet introducing a totally new breed of animals on earth. Some scientists believe what caused acanthostega to move from the sea to dry land may have been the evolution of predatory placoderms like dunkleosteus. Dunkleosteus was a nasty looking beast and one of earth's first major predators. This formidable character would have made living in the sea less inviting for animals like acanthostega. One day, when the sun came up, acanthostega was running (literally).

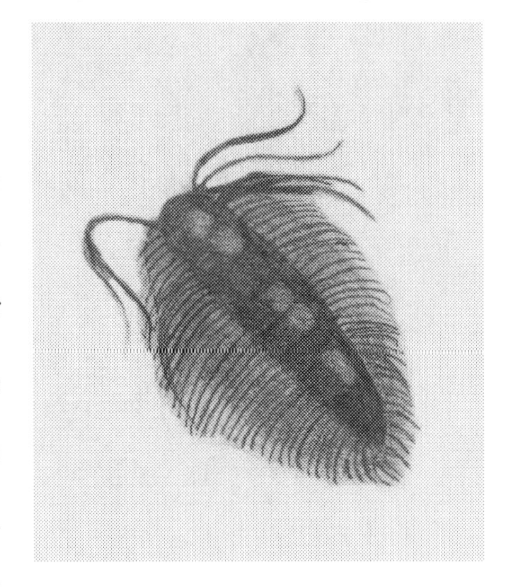

The Carboniferous Era brought some of earth's most dramatic animal and environmental changes on land, sea, and in the air with its high oxygen levels and abundant fauna. Earth's fossil fuels (coal, oil, gas) were all produced during the Carboniferous Era. Animal life, including some monster insects flourished during this 200 million year time frame. Earth's evolutionary machine marched onward to the giant dinosaurs of the Jurassic period which were followed by a totally new variety of life, like the titanis bird, in the Createaceous and Teritiary periods. Eventually planet earth somehow managed to end up with human life forms as evolution produced our current living world. It has been a process of constant change, modification, and evolution. Despite a history that included six major waves of animal extinction, life on earth continues.

Life form changes had been gradual yet monumental as our earth remodeled itself over and over. Overtime significant animal adaptations (like legs for acanthostega) had to be generated on a never ending basis for life to continue. Let me take a look at a few interesting animal adaptations that took place, ones we are all familiar with.

Interesting animal adaptations

Early earth history gave us examples through fossilized remains of creatures like acanthostega, telling the story of adapting/evolving from life in the sea to life on land. Fossils also helped to reveal a few of the billions of major life form changes during the Carboniferous Era 200 million years ago including enormous insects, like dragon flys with three foot wings. Later the Triassic and Jurassic periods manufactured mega dinosaur monsters, like tyrannosaurus rex, brontosaurus, triceratops, and so many others that we may never discover but will forever engage our imagination.

Even the giant tyrannosaurus rex evolved from a four legged beast to one that stood on its hind legs with small forearms. The T-Rex forearms were rendered useless, no longer needed and, over time, atrophied into two small unused legs that would have eventually disappeared. Birds, like titanis, mammals, and sea life following the dinosaur demise were truly impressive animals that have been overshadowed by the giant dinosaurs preceding them and ruling earth for millions of years.

Adaptations can take a number of different forms in body and behavior. Camouflage coloring was developed to conceal predators or protect prey. Animal structure, feet, hooves, claws, or teeth did change to meet defense or food gathering requirements. Some significant body parts were simply not used over time such as the wings on the, now extinct, Do Do bird, wings on present day ostriches, or the wings on penguins that are not used for flying but evolved to propel them in the water. Behavioral modifications had also been made to meet environmental changes such as colder/hotter temperatures or disastrous events such as ice ages, tectonic plate movement, volcanic eruptions, mega fire storms, or floods that would have permanently altered the landscape.

Early evolutionary science

No book written regarding evolutionary thought would be complete without reference to research done by Great Britain's naturalists Charles Darwin (1809-1882) and Alfred Wallace (1823-1913). They were the co-discoverers regarding the theory of natural selection in evolution. Their research and the discoveries they made in the mid to late 1800's had a profound impact on modern understanding of life on earth and animal development. They lead the way regarding present day understanding and acceptance of evolution. Darwin and Wallace identified the gradual process by which biological traits become either more or less common in a population. Isolated islands in the Pacific and Indian oceans (Galapagos Islands for Darwin and Indonesia for Wallace) were ideal laboratories used by them to identify secluded animal changes taking place over time. Darwin discovered thirteen finch species that had developed totally different beaks even though all of them evolved from a common ancestor. Each beak, large, small, narrow, wide, all evolved to meet the food available for the finches in each secluded island area.

Their research while revolutionary (and correct) was not met with great favor outside the scientific community, certainly not with religious creationist idealists. In religious circles the work done by Darwin and Wallace was received much like the work of Galileo Galilei when he submitted that the earth was not the center of our universe. Unlike Galileo however Darwin and Wallace were not found guilty of heresy, imprisoned then placed under house arrest for life. Times had certainly changed and shocking new scientific research had been found more acceptable.

Life form and behavioral changes on our delicate planet have been and always will be a slow evolutionary process. Animals that met a changing/evolving environment were the ones that lived to pass on their genes to future generations, animals that did not simply vanished.

Examples of this evolutionary process are visible everywhere. Look around at the animal variations, adaptations in your yard or a park nearby. Evolutionary examples are present and underway everywhere, a turtle, a deer, a squirrel, or fish in a pond are great examples of evolution/adaptation. You have to wonder why they became the animals they are and how did they acquire the unique hunting, feeding, and living tools that they employ each day. To succeed in this changing world all animals must adapt or perish, simple as that.

Evolution of Life on Earth

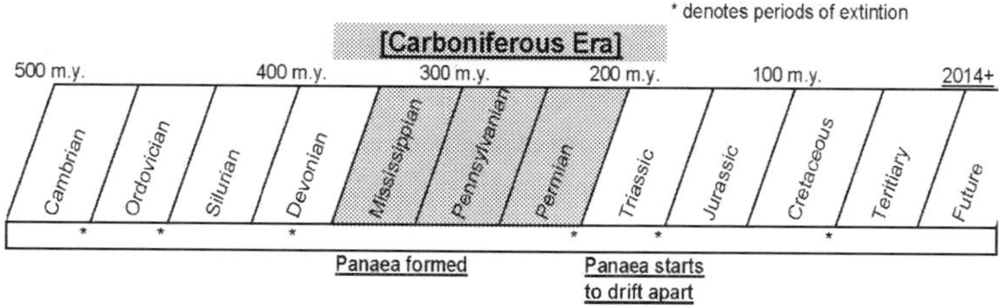

* denotes periods of extintion

[Carboniferous Era]

500 m.y.
400 m.y.
300 m.y.
200 m.y.
100 m.y.
2014+

Cambrian *
Ordovician *
Silurian
Devonian *
Mississippian
Pennsylvanian
Permian *
Triassic *
Jurassic
Cretaceous *
Teritiary
Future

Panaea formed

Panaea starts to drift apart

Chapter Two

The insect world

What "bugs" me

(Fire ants and killer bees)

Let me begin with my thoughts on the smallest, yet potentially the most dangerous, of wild animals, the insects, those little guys that bug us all. Insects interact with humans every single day and, small as they are, insects may well be the most voracious and persistent predators humans are up against.

Insects represent over 80 percent of all life forms on earth and simply cannot be avoided. From nuisance pests to massive swarms of locusts, the insect world is on a constant collision course with ours. The magnitude of this conflict can grow exponentially, becoming more complex as bugs continue to evolve and adapt.

In college, an entomology student resided in an adjacent dormitory and this character was considered something of an oddball in dress, mannerisms, and his outlook on life. His small stature and nuisance personality earned him the nickname "mosquito" by his roommates, and I honestly don't recall his name. My friends and I simply referred to him as the "Bug-Boy." While he was a bizarre character, I do remember him as very bright and interesting to talk to, as often he described insects in a way that made them real, gave them feeling, direction, and personality. He was fond of telling anyone who would listen: "Insects rule, humans don't." His mantra. Reflecting back on it, what he said may be true. Let's look at a small sample of ants and bees that exemplify Bug-Boy's belief.

Ants

Ants are incredible animals, and they are everywhere. Living in a wooded area, I cannot seem to keep those pesky ants out of my house. A neighbor recently commiserated with me, relaying on-going problems he had with ants getting into his house. We were standing in his driveway; he was stepping on ants all around him and looked comical as if he was dancing the cucaracha.

He said "I'm pulverizing these little bastards all the time, and yet they keep coming." I told him he could have David Flatly and his flaming-feet Irish dance ensemble running up and down his driveway, and he still would lose his war with these ants. Even chemical poisons can only deter or temporarily slow them down, but that's all. Those determined little bugs always return.

Ants are phenomenal animals, and if you desire to see dramatic life and death struggles first hand, up close and personal, just take time to watch ants marching through your backyard, garden, or in a park near your home. Ants are constantly on the move, they have excellent communications, they are tireless workers, they fight like demons, and they never appear to rest.

Fire Ants

Introduced "accidentally" in the USA by a South American cargo ship around 1930, fire ants have spread rapidly throughout the Southern states as well as the Southwest. They have now traveled as far north as Kansas and Maryland. Fire ants are social animals and very organized. Each ant has a job to do every single day, and it never stops (no weekend breaks or vacations for these guys). Some fire ant colonies hold 100,000 to 500,000 ants. Fire ants are omnivores and prefer protein foods

(other insects and meat), but will eat almost anything. They are very aggressive and one of the world's strongest animals, as they can lift items fives times larger and heavier than themselves.

Tragically, fire ants have now been found on one of the Galapagos Islands. For centuries these isolated islands served as the bastions of evolutionary evidence examined by scientists from the time of Darwin forward. Now, thanks to human intrusion, that has certainly changed.

Fire ants are tough little customers, and unlike other ants, only bite to get a grip on their victim. With the grip in place, they use a stinger in their abdomen to inject a toxic alkaloid venom. It is a painful sting that causes a burning sensation, which is how they got the name "fire" ant. These guys can sting repeatedly. One sting is bad; hundreds are horrible and can be fatal.

Fire ants continue to evolve and expand their colonies, adjusting, adapting and moving north. So far efforts to exterminate or control them have only been met with limited success. Bug-Boy said these guys are unstoppable, and he may have been right.

The US Federal Drug Administration estimates that over five billion dollars is spent annually for medical treatment and control in fire ant infested areas, and fire ants cause nearly one billion dollars annually in agriculture, livestock, and crop damage. Hey, that's a lot of money for some tiny little bugs!

Fire ants continue to adapt and evolve adjusting to colder climates as they move further and further north.

Killer Bees

Killer bees were introduced into Brazil and parts of South America to replace European honey bees. European bees were preferred, however they did not thrive in South America because they had a hard time adapting to the climatic conditions. Killer bees were brought in to take the place of European bees, and since their introduction killer bees have expanded at the incredible rate of two hundred miles a year across South America, then Central America, and now into the USA. They reached South Texas in 1990, then to Arizona in 1993, and California in 1995. They are now spreading fast throughout the Southeast USA, moving as far north as the Chesapeake Bay. Like fire ants, these little bugs cover great distances and seem unstoppable.

Killer bees are highly defensive and very aggressive. They will attack in mass even when unprovoked and will respond rapidly to any disturbance to

their nest or invasion of their living area. They have a single sting, and like European bees, die after stinging their victim. Killer bee attacks are relentless and in large numbers. They can sense human threats fifty feet from the nest and will pursue an enemy two miles or more. If a person being chased by these bees elects to dive under water to escape, the swim will do little good, as these war-like bees will wait for the human to come up for air. They are notoriously aggressive and determined animals. Killer bees have been credited with the deaths of more than one thousand humans since being introduced in Brazil.

Typical killer bee colonies house eight to eighty thousand bees, with one colony discovered in Bisbee, Arizona, harboring two hundred fifty thousand bees. Here too, attempts to exterminate or control killer bees have only been met with limited success.

Killer bee control costs soar into the billions annually, and yet they continue to flourish and expand their territory (evolve/adapt).

Something to think about

The Cleveland Museum of Natural History recently had a Tyrannosaurus Rex exhibit on display. This exhibit was on loan to the museum from the Field Museum in Chicago, and the display was ever so impressive. To see that monster up close you realize how helpless or hopeless it would have been for humans to reside in the Jurassic period.

At this same exhibit there was another animal that impressed me even more than the T-Rex and that was a life-size display of a mosquito from the Carboniferous era. This mosquito had a body the size of a man's fist and could certainly draw more blood from a human than a Red Cross Nurse at the Blood Bank. A swarm of these monsters would leave a trail of exsanguinated human corpses everywhere they went. You wouldn't need the T-Rex to do in humans—the bugs would put an end to humans fast.

Granted, these bugs lived in the Carboniferous era, when the earth's oxygen levels were far greater than today. With no lungs, oxygen content in the atmosphere controls insect growth. With more oxygen you got bigger bugs, conversely, with less oxygen they were smaller. A dragonfly from the Carboniferous era would have had a three foot wingspan. Just think about that for a minute. Dragonflys are voracious predators, fast, furious, ruthless in their attack, and can stop or turn in an instant, fly backwards, sideways or any direction to catch prey. Watch one at a pond near where you live, and you will see one nasty, aggressive bug. Those guys wake up running.

The 30 percent oxygen levels of the Carboniferous era are not likely to return soon, but what if there was a slight environmental change, and it went to the bug's advantage? A dragon fly today (with 20 percent oxygen levels) has a wing span of around three inches. If a 50 percent increase in oxygen levels produced three foot wings, what would a ten percent change produce. How would you fight off fire ants or killer bees that were three or four times as large as the one's we have today? A squad of jumbo ants or bees would be enough to take out a human. A swarm of a few thousand could likely take out a neighborhood.

Several years ago many "B" rated scare-movies were produced that portrayed larger than life insects attacking humans. Humans were helpless against those giant bugs that systematically tore them apart. Armed forces, guns, cannons, and bombs did not stop those monster movie bugs. Were those old movies a harbinger of things to come? Are climate changes and global warming working to insects' advantage? It's certainly something to think about.

This chapter only briefly examined our relationship with bugs, and I selected two insects that have been in the news that readers are familiar with . . . didn't mention the monster problems that have historically complicated human existence by mosquitoes, cockroaches, or locust swarms. A miniscule size change here would wreck havoc on humans as some past locust swarms were estimated to be miles wide and long, devouring everything in their path. The simple fact is that there are millions of different kinds of bugs and trillions of bug animals that call earth home. Humans share space with bugs, and at times there isn't enough space. My presentation is made to get readers thinking. Thinking about the small changes and thinking about the large ones.

Question

Based on this abbreviated discussion, the bottom line question here has to be this: Do you want a nest of fire ants or killer bees in your backyard? Not this writer. However, in time, they will be there. You see, Bug-Boy may have been right.

Chapter Three

Climb up the food ladder

Birds of prey

(True decedents of Tyrannosaurs Rex)

Birds represent the one animal group I highlight that do not currently present a direct threat to humans. They do however interact with humans every day and have changed how they view and deal with their human menace.

Birds of prey are truly impressive animals. Unlike insects, birds have the most efficient blood/oxygen system of any animal, humans included. Birds of prey have refined this energy enriching system to enable them to fly and soar for hours at a time. They can climb to and perch high in a tree, on a building, or any elevated structure and spot the slightest movement on the ground. They are fine-tuned hunters with excellent flying skills, powerful talons, terrific speed, and binocular vision (a quadruple threat). Were these animals larger in size, say the size of a pteranodon, humans would be no match and soon end up as bird feed.

With human intrusion into the wild, coupled with wild intrusion into the urban world, conflicts do occur with costs to both sides. Household pets are often the victims because, when other food sources dry up, our fat little over-fed household pets become an attractive food choice for birds of prey.

The dictionary defines predator as: "One that preys, destroys, or devours an animal, one that lives by predation." Predation is defines as: "A mode of life in which food is primarily obtained by the killing and consuming of animals." Humans certainly fall into this category, as carnivores killing animals with bullets or dollars, for food.

Birds of prey fit this definition to a "T" and I submit there have been numerous reports regarding human pets lost to eagles, hawks, and owls. Most of these losses are truly heart-felt losses, as domestic pets become devoted family members, providing unqualified love and friendship. Losing a pet in an instant strike from the air leaves behind a crushing memory. To a bird of prey, a small pet running through a backyard is simply easy picking and a juicy dinner, no more. Compared to catching a squirrel in the forest, these opportunities humans make available represent a rewarding environmental transition for hawks, eagles, and owls.

Birds of prey are also territorial, driving off any intruders, humans included. Once, in preparation for deer hunting season, I located an excellent spot to build a tree hunting stand. That stand would provide better hunting visibility and greater stealth (silence). Prior to the season I had my stand in place and did not return to the area until just before sunup on the first morning of hunting season. I stayed away so the site was not contaminated with fresh human odor. What I didn't know was my hunting location intruded on the hunting area claimed by a large owl. This owl did not appreciate my intrusion, and as soon as I entered my stand in the early morning darkness of the first day of deer season, this owl let me know it. The owl repeatedly dove directly at me, hooting and screeching. I felt like I was in Alfred Hitchcock's movie *The Birds* with this big owl making repeated, determined dives directly at his target, me (truth is, this guy was scary). Needless to say, no deer would come anywhere near my hunting stand with this chaos underway. I had no choice but to abandon that area and departed with the score: Owl 1, Bill 0.

Birds are intelligent animals, too. Examples of bird intelligence can be seen on any day if you just watch the crows or ravens in your neighborhood. The dictionary defines a crow as: "Large glossy black, oscine bird and member of the corvidae family" and a raven as "a glossy black corvine bird related to the crow family. Hey, right or wrong, I call them all crows.

Studies have proven that crows can count, can reason, and are not intimidated by human built "scare" crows. They calculate intervals farmers set on carbide cannons (designed to explode periodically, driving crows away) and while adroit at avoiding unwanted human contact they reside everywhere from the parched deserts of the southwest to the Arctic North Slope of Alaska.

Once driving down a rural highway, a flock of crows were gathered several hundred yards ahead, scavenging a road killed carcass on the highway. A back seat driver in the car, concerned that I might hit the crows, cautioned me to slow down. Knowing crows and being a tad obstinate regarding back seat driving advise, I hit the throttle, not the brake, aiming at the crows in what this person thought was an act of pure madness and described as a fiendish, deliberate attempt on my part to massacre an innocent flock of crows. Now what do you think really happened? Aside from the loud screaming in the car, those crows, in their cool leisurely way, gauged my speed, lifted off the highway and out of harms way, returning to their road kill dinner as soon as danger passed. I submit that you cannot hit those guys on the highway, because they are simply too intelligent, too cautious, and, when hunting, operate a step ahead of humans. Roadsides are littered with the carcasses of a variety of animals: raccoons, possums, skunks, deer, and an occasional bird of prey, but never crows. I tell you those guys are clever; they wake up running.

Around a campfire in Central Alaska a few years back, Patrick Wheeler, a registered and licensed Alaska hunting guide, and I were talking about the wisdom and cunning nature of crows. We shared stories about crows being able to count the number of hunters going into and out of cornfield blinds and how they call out warnings to near by crows and other animals. One story involved crows perched on a snow-covered roof ledge above a building door waiting for humans to exit. When any unsuspecting human opened the door and walked out, the crows would kick snow down on his or her head, apparently for sheer entertainment. Those caw caw caw's you hear as they fly away are sometimes crows laughing, having fun, and being ornery at the human's expense. Translating those caws into English might be: "Hey, Charlie did you see me nail that dummy?"

Patrick also shared several stories about ravens following Alaska sheep hunters. Here hunters are rightfully concerned that the noisy, circling crows might give away their position (has happened several times to me when deer hunting). These clever crows know when the human sheep hunter is successful; the gut pile left behind (for crows, the best part) provides a fantastic dinner. Crow's understand that it's in their best interest for the hunters to succeed!

Let me give you this example: Hunting sheep in the mountains of Alaska is very physically demanding. Sheep live at high altitude (thin air), on mountains with steep rocky inclines, often in open areas where predators (wolves, bears, humans) are visible and easily avoided. Getting close enough for a good rifle shot can be more than mildly difficult and often impossible. I heard of an instance where crows were actually seen diving at a herd of sheep, trying to push the sheep closer to the hunters for a shot. On this day the hunters had stopped their stalk as they simply couldn't get close enough for a clear shot at a good ram. Then, unbelievably, the ravens decided to help the hunters, and it worked. Diving at the sheep, the crows managed to push a nice ram into shooting range for the hunters. After the hunters finished field dressing the ram and packed out the meat and trophy horns, those crows flew in to enjoy a well deserved dinner, feeding on the animal "gut pile" left behind. Crows are amazing animals—very skilled, very smart.

Birds of prey are excellent hunters and can hinder (in some cases help) human hunters in the wild. They hinder a hunter by the warning noises they make when a hunter moves through their territory (their home). The racket they make when humans invade "their" territory (hawks, owls, and crows) is phenomenal, sending warning signals to every other wild animal in the area.

Think about this

What if evolution suddenly accelerated and birds of prey became much larger? Today the monster raptor is the Stellar's Eagle at 20 pounds and a seven foot wingspan. In the Cretaceous period the pteranodon had a wingspan of twenty feet, and as mentioned earlier, a prehistoric dragonfly from the Carboniferous era had a wingspan of three feet. Another monster to think about that lived closer to our human existence (approximately three million years ago) was the titanis bird that evolved in South America and moved north. This flightless carnivorous bird was ten feet tall, could run at forty miles per hour, and looked like a tyrannosaurus-rex with feathers. Titanis was no park pigeon looking for a popcorn hand-out. It hunted like an aggressive tyrannosaurus-rex, and I submit would view humans as "easy-picking." When the sun was up, this guy could run.

A flying example of the titanis monster would be the teratorn, found in the La Bera Tar Pits. This giant bird had a sixteen to twenty foot wingspan making it close to the size of a small airplane and certainly not a bird you want taking roost on your house. Some scientists do question the diet of the teratorn, that it might have been a scavenger like the California Condor. Other's in the science community believe it may have been an omnivore or pure carnivore, if so, I suspect it would see humans as dinner fair.

If an eagle, owl, or hawk had an enormous twenty foot wingspan or gained the size of Titanis, things would get ugly fast for humans.

The dramatic evolutionary change I reflect on is not likely to happen in our human lifetime as our total existence on earth is a mere blink of an eye in cosmic or geologic time. However tiny, incremental changes, components of a bigger picture, may be taking place right before our eyes.

Raptors may be part of the new order.

Chapter Four

Coyotes and Wolves

(Yes, they sometimes attack and kill humans)

Coyotes and wolves rank high in human interest—very high. They are intelligent, strong, resourceful, and gifted animals. Few people hold a "middle-of-the-road" opinion of them, and the positions often rest at the extreme; people adore and worship them, or people want them eliminated. Love them or hate them, it's a heated debate that has no middle ground. Regardless, it is important to respect coyotes and wolves for the resourceful animals they have evolved to be.

Taking a position supporting either side makes you an instant enemy of the other, and it was and is my desired objective not to join either side of this unending debate. Through my observations in the wild I simply took a brief look at where these animals lived, how they hunted (or were hunted), and how they interacted with humans. My plan was to uncover any changes that might put coyotes and wolves into the new-order scenario.

Coyotes

These impressive animals can be found from Panama, through Mexico, the USA, into Canada, and most of Alaska. With nineteen subspecies, coyotes evolved in North America to fill an ecological void by preying on small mammals that, left unchecked, would over-populate the habitat, causing mass starvations and environmental chaos.

Animals coyotes focus on to eat (mice, rabbits, gophers) are prolific, not too bright little guys who have no concept of longevity. They make easy targets for carnivores residing just a step above them on the food ladder. No retirement plan needed for these guys, they just eat, sleep, and reproduce, some delivering litters three times a year. Populations left unchecked by predators like coyotes would explode overnight. Coyotes stepped in, did their job, and did it very well.

Coyotes weigh up to 50 pounds, can run at over 40 miles per hour, and may reach 50 inches in length, head to tail. They are clever and adapt well to any new environment.

While generally solitary nocturnal hunters who prefer avoiding human contact to pursue small prey such as mice or rabbits, coyotes have joined forces forming packs to take down larger prey. Coyotes have strong endurance, and no matter how they hunt (pairs, packs, or solo), they are very persistent, intelligent, and determined hunters. Maintaining communications and working as a team, coyotes run deer to exhaustion or drive prey to a hidden member of the pack, lying in wait (similar to crows helping sheep hunters). These skilled hunters have teamed up to kill animals such as elk that are ten to twenty times their size. In some areas of North America, deer fawn deaths have been reported to be as high as 80 percent with coyotes listed as the number one villain.

A game management official once told me that he would rather take on an angry 150 pound German Shepard dog than a 40 pound coyote. He had been called on to remove "problem" coyotes in the past and learned through his trapping experiences that coyotes can be very tough customers. Just as the saying goes: "It's not the size of the dog in the fight, but the size of the fight in the dog."

Livestock farmers consider coyotes a true monster. Coyote packs expand their territory and too often develop a taste for domestic sheep at dinner time. Those fat, slow sheep are much easier to stalk and take down than a big, fast elk or full grown deer after fawn-birthing season.

While coyotes are wild beasts, they also do well in urban settings. In New York City, coyotes roam "free" in city parks and often venture into the city itself. In March of 2010 NYC Police reported a coyote running through Manhattan (the fourth coyote cited year-to-date) that made big news. NYC Coyote number four became known as "Wily Coyote" to residents, and after many sightings of Wily, it still took days before he was finally captured. NYC Officials then stated that they planned to release Wily into one of the city parks, conversely Wily Coyote was probably planning when to make his next run through Manhattan. Cities across the United States and Canada had news reports of coyotes invading their city limits, living comfortably with their cautious nocturnal travel and easy human food offerings. The new coyote menu included: household pets, park animals, and fast food dumpsters.

Sadly, reports of some coyote appearances came too late. There was a news article in May of 2008, published after a two-year-old California girl was attacked and pulled from a public park by a coyote. The girl was saved but suffered bite wounds and, certainly, emotional wounds as well. This article reported there had been over 100 coyote attacks on humans in Southern California since the 1970s. All the reported attacks were made on children under ten years of age, with one fatality in the City of Glendale. Just like hunting other big game, coyotes go for the young, the sick, or the wounded first.

Another tragedy occurred in October of 2009, when a young Canadian singer, songwriter named Taylor Mitchell was killed by coyotes. She was hiking solo in a Canadian National Park when she was attacked and killed by a pack of coyotes. Sad not just for the loss of life, but the attack was brutal and happened in broad daylight—so unusual for a pack of coyotes. This coyote pack deliberately stalked, attacked, and claimed the life of a young adult.

While attacks on humans are rare, they do happen. Is this a sign of the new order? Do these new attacks represent a microscopic component of overall evolution on earth?

Wolves

Unlike killer bees or fire ants, wolves shared hunting land with humans from the start of their predator relationship and were always prime candidates for hate, fear, and distrust. Early (now extinct) dire wolves to present day gray wolves were always viewed with trepidation, suspicion, and as mortal threats to deal with. Wolf habits and lifestyle were so secret that they easily became mythological monsters. From before the days of Little Red Riding Hood wolves were victims of misunderstanding, folk-lore, hatred,

and fear. Let me add that part of historical dislike or hatred was caused by wolves being much better hunters (competitors) than their human counterparts (wolves wake up running).

A wolf is certainly an animal to fear. Present day male wolves can weigh over 150 pounds, they have tremendous stamina, and can travel hundreds of miles over unforgiving terrain. They are highly intelligent, canny, pack hunters that communicate with each other to accomplish group objectives. A wolf's predatory stare alone is something that could send a chill up a hardened hunter's spine.

Wolves were (and remain) difficult animals to study. Some researchers have done remarkable work in this area, and let me mention just a few: First, I consider the best all-time wolf study to be that conducted by Farley Mowat and presented in his book *Never Cry Wolf.* His book was later made into an entertaining and educational movie. Here Mowat was able to capture wolves in their natural habitat and uncover many truths about the animals, placing wolves in a much more favorable light. While his mission as a Canadian government biologist was to uncover how wolves were the vicious animals responsible for devastating herds of caribou, he found just the opposite. Wolves had a varied appetite depending on the season. Their food was small and large. Wolves ate mice, ground squirrels, even fish, along with large ovulates. He further determined that wolves helped strengthen caribou herds by removing the frail, sick, old animals, making the herds much stronger. Bad weather and over hunting were more the culprits in the decline in caribou numbers. Mowat's research also took place in the arctic where he followed, and scientifically studied, a pack of wolves in their natural habitat, a monumentally difficult undertaking. Findings in his study went against the wisdom of the day, and his research was not viewed with favor or fully appreciated. Mowat's investigation however was an excellent first-hand examination that revealed many new truths about wolves, and his book is an excellent read.

A second study conducted by Jim and Jamie Dutcher was designed as, and turned into, a film documentary and book, *Wolves at our Door.* This six-year study revealed much more about the complex social structure of a wolf pack and what magnificent animals they were, earning a meaningful place in our eco-system.

The Dutchers' study did have limits that they were honest about. The wolves studied were not fully wild animals, as some were human-raised. They were also human fed during this study and were restricted to a twenty-five acre enclosure in Idaho (not much territory for an animal that can travel hundreds of miles). The Dutchers' did an excellent job depicting the complex social structure maintained in a wolf pack.

The anti-wolf groups (hunters, hunting guides, ranchers, farmers, and other Little Red Riding Hood types) claim wolves actually do more harm than good to ovulate herds, and wolf numbers need to be better controlled (or wolves banished). A few also submit that wolves engage in surplus killing, or killing ovulates to train their young, killing for practice, or just killing for fun and wasting precious game animals. Another claim that angers anti-wolf groups is that the wolves' killing practice was not pretty. Wolves were designed as apex predators but fall short regarding skills needed to take down large game. Wolves, with their great endurance, run a deer to exhaustion, then one wolf would hold the deer in place by biting and clamping down on it's nose while other pack members tore into the sides of the victim, literally eating it alive. Wolves are not always nice guys.

These anti-wolf groups believe that wolves are not needed for herd control, and some pointed to New Zealand as an example. New Zealand has no natural large predators yet maintain great herds of ovulates such as: red deer, elk, fallow deer, tahr, and chamois. These herds are controlled through licensed hunting and formal culling programs when needed (big game hunting in New Zealand is big business).

Donald Benedik, a good friend of mine and very experienced hunter, shared with me information he was told by an outdoor guide he met while hunting in New Zealand. This guide had previously worked for New Zealand's game management department where his job held some responsibility for herd control. The government there strictly regulated hunting and culling programs. One year the game control department was called on to shoot over 1,500 red stags by using helicopters as a hunting platform (a significant culling plan). The red stag carcasses were transported to packers and butchers where most of the meat was sent to Europe for restaurant consumption. Right or wrong, the score here was: Restaurants 1,500, Wolves 0.

Hate or love. Sides periodically change in this debate, as through education, a few wolf-haters were turned into wolf-admirers, and a few others, who experienced unfavorable wolf encounters, went the other way. There still could be little or no middle ground. The strength of emotion runs deep, and those extreme postures rarely go away.

Truly, if coyotes had no middle ground in the love-hate battle, wolves clearly took that battle to a much higher level. Wolves marched into an unending and unfavorable limelight, because the human population perceived them as dangerous villains, terrorizing communities, and killing any humans standing in their way. Perception will always override fact.

Let me share two personal encounters I had with wolves in central Alaska. One encounter was favorable and one not so favorable.

My favorable encounter took place during a moose hunt in the Wood River area of central Alaska. A pack of wolves were seen on a distant ridge, running the ridge line and working back and forth on the side of a small mountain. I trained my binoculars on this pack and watched them as best I could, but the ground cover was high and the wolves were in the open just a short time. What they were hunting, I could not see and suspected the wolves were working as a team to drive ground squirrels or whatever toward other members of the pack. There were crows in the sky, and my hunting guide advised me that often wolves follow crows (or vise versa), because crows will locate carrion that they are unable to eat as the animal hides were too thick. By drawing wolves to the scene, wolves would tear apart the carcass and, when finished, would leave behind a nice meal for the crows. Interesting too is that those crows might have been working to drive an animal toward the wolf pack. I sat there watching the crows overhead and these wolves running back and forth on the ridge and thought maybe they were in communications with the each other. Both animals are extremely intelligent and have strong communications skills, so maybe, just maybe, they communicated with each other organizing this hunt. Scary, wolves hunting with spy drones, like Army helicopters supporting M-1 tanks. A tough combination.

On this day, we humans were hunting our game in our area, and wolves (with crows) were hunting their game in their area, no connection between us. I however strongly believe that if our human hunting party would have moved closer to the wolf pack, the wolves would have vanished into the wild long before we arrived. This encounter represented what the vast majority of wolf encounters had been: Wolves did not want human intrusion and maintained a safe distance away

My second encounter was different. I was on a guided moose hunt in central Alaska, and it seemed that my "guide" was determined to turn the hunt into the hunt from hell. It turned out this person (let me call him Frank) was not a licensed Alaska guide at all but worked with two other individuals who were guides. For some reason, unknown to me, the two guides turned me over to Frank for this hunt and then went off in another direction with a second hunter. Frank really did not know where he was or where to go. It was a classic case of the blind leading the blind.

The first day out was a long one where we covered a tremendous area, walking non-stop for hours. The tundra area of Alaska and the mountains were killers on my knees. Unless you have been through this country on foot you cannot appreciate how difficult the travel is on foot. It was a killer. The soft mud and swampy tussocks combined with the spongy moss covered hillsides just took a toll on legs . . . three steps forward and one step back (all day).

At the end of this day, we headed back, attempting to locate our camp (we were lost more than Frank cared to admit). It was dark, and we were long overdue. Eventually we came upon Virginia Creek, which I knew from my map of the area would eventually lead down stream to the Wood River and then to camp. Going down Virginia Creek, however, was a monumental task. With my knees failing, it was all I could do to keep upright and moving, but Frank was nowhere to be seen. Virginia Creek wove through the woods getting narrow at times and other times it widened out, shallow then deep, and the banks along the creek were very thick with vegetation. Getting off the creek to avoid a deep area was nearly impossible, and at one point I decided to cross a narrow stretch I thought was still shallow enough that I could get through. I was wrong. It is not possible to describe the chill of that cold water that filled my boots. My legs were shot at this point, I was soaked, and I was in trouble, barely able to make headway down the creek.

Frank had his own agenda that day, and it had nothing to do with guiding me. I rarely saw this "guide" throughout the day. Now with the pitch dark upon us, he dropped back a few times to tell me: "We have to get off this creek; the wolves run the creek at night." Then Frank disappeared. I had heard wolves howl the night before and had no reason to doubt they were near. I was in a fix because I was the injured, trailing animal and the one a predator would focus on. Eventually, I made it to the hunting camp (tents), but it was 11:30 at night. I was soaked, in pain, and totally exhausted.

The next day I insisted on travel using the horses (that I was paying for), as walking any distance on day two was out of the question. I cut myself a walking stick and was able to get around okay, but to travel all day on foot, in that unforgiving terrain, was not in the cards.

Riding up Virginia Creek that morning I came across wolf tracks, some big ones, in the mud and sand adjacent to and on top of my boot tracks. When I think about that encounter, to this day, the dark of night, stumbling down a strange creek in Central Alaska, with a wolf pack trailing me, it sends a chill up my spine. I was lucky.

Wolf attacks on humans

I have often been told that wolves do not attack humans and there were no recorded attacks on humans in North America. That is something I told myself over and over as I stumbled down Virginia Creek on that late dark night in 2006. There was, however, a wolf attack I had been told about that happened only the summer before my hunt. A bicyclist on Highway 3, the George Parks Highway, near the Denali Park, had been chased by a wolf, but the encounter was not bloody.

In December 2007, *Alaska Magazine* reported that there had been three wolf attacks on humans the prior July. None of the three were fatal, and each one was a single wolf attacking a human. First a hiker, then a motorcyclist, and the last attack made on a bicyclist. The hiker suffered bite injuries before she reached safety (an outhouse), but the motorcyclist was able to out run the wolf, and the bicyclist encounter ended when the wolf was driven off by a truck headed the opposite direction on the road. The truck driver saw what was happening and was able to steer into the wolf's line of attack, driving it away. Fortunate for the bike rider because that wolf could have easily out run him on the bicycle, and the rider had no other defense other than the speed he could make on a bike. Not fast enough.

American Hunter Magazine reported a wolf attack made on an elk hunter named Rene Anderson in 2001. Fortunate for Rene, she was armed and able to kill the wolf, although it came within ten feet of her when it charged.

There have been other recent, non-deadly, human/wolf encounters in Alaska.

A few recent encounters making the news have unfortunately been deadly. In June of 2012, a Swedish zoo worker was killer by wolves when she entered the wolf pen. Here workers often entered the wolf pen alone, and the cause of this attack remains a mystery. It stands as testimony that wolves are truly wild animals and, caged or not, can turn on humans.

In March of 2010, Jogger Candice Berner was attacked and killed by a pack of wolves near Chignik Lake, Alaska. Candice had only recently arrived in Alaska to start a teaching job and, alone, was defenseless against the wolf pack.

When a wolf pack launches an attack on an animal (humans included), they have one intention. To kill.

Change underway?

Are these attacks just an anomaly, just happenstance, bad luck, or is there a microscopic change underway in the man and wolf relationship and earth's evolution? Do wolves see humans in a different light? Clearly, if wolf attacks didn't occur in the past, that scenario has now changed. Might this be a clue?

Chapter Five

Cougars

True Ambush Hunters

(They too can have an appetite for human blood)

Cougars are truly mysterious predators, so mysterious that I had to include them in my thoughts on animal change and evolution. These impressive cats go by several names: cougar, puma, mountain lion, or panther, but they are all the same breed of cat. This animal is so private and mysterious that it has always been hard to evaluate, and very few accurate studies have been made over the years, so few that recent attacks on humans may not represent the behavioral change I now submit. Looking far back in human history, the cougar's relative, old "saber-tooth," had to have held a position of fear in Neanderthal and/or Cro-Magnon eyes, and I suspect this cat never truly held humans in high regard, just easy prey and a tasty dinner when needed. Maybe yes, maybe no.

A cougar's range can cover 300 to 500 square miles, and even though it was once hunted to near extinction, its territory is now spreading, reclaiming areas in the West and moving east. Verified cougar sightings have been made in Illinois, Indiana, Michigan, and it has even turned up in areas of New England where it was driven away decades ago. Standing thirty-five inches tall and 8 foot from head to tail, males can weigh on average 150 pounds. Some cats are reported to reach over 220 pounds. With rear legs longer than front, these cats can leap the length of a school bus (over 18 feet) and run at 40 to 50 miles per hour. As with most cats, the cougar is not a distance runner but is fast and agile, using it's long tail for balance in rough, uneven terrain. These cats are the epitome of stealth, grace, and power.

Ungulates represent the bulk of a cougar's diet, and when deer, elk, or sheep are hard to find, the cat will turn to domestic animals on ranches, farms, or in your backyard. It's no problem for a cougar to take down quarry that is seven times its size, and it can be a voracious eater, killing, on average, one deer every week to ten days to fill its appetite. Horses, cows, sheep, and domestic pets are easy prey for the mountain lion and can represent a tasty diet change. Sadly, Cougars have also reportedly engaged in surplus killing with one rancher losing twenty sheep in a single attack.

It is a territorial and reclusive nocturnal hunter that excels at stalking and ambushing prey. A cougar's attack is sudden, unexpected, and deadly. While it prefers to hunt at night, it can also be a diurnal or corpuscular hunter. It adapts to change well, including human invasion and expansion.

Raptors in North America have a hunting success rate of one kill in every ten hunting attempts (ten percent). Human moose hunters in the Great State of Alaska have a success rate of less than ten percent. However the cougar is just the opposite, with a hunting success rate in nine out of ten attacks. A 90 percent success rate makes these cats a top predator and one of the best hunters in the animal world (humans included). When the sun comes up, these guys are running.

I have had only one wilderness encounter with a cougar, albeit a panther. In Florida, on a trail near the Everglades a panther jumped out onto the trail about fifty feet in front of me. This was in broad daylight, and the cat was as surprised to see me as I was to see it. I have no idea what caused this cat to be on the move in the middle of the day, however, upon seeing me, it instantly turned, jumping back into the thick cover of the swamp, and vanished from sight. Not all human-lion encounters turn out as favorable as mine. In Florida, this cat was not stalking me and wanted nothing to do with me (or me with him). However in Colorado recently a mountain bike rider was ambushed and killed by a mountain lion. I suspect this guy never knew what hit him when attacked by the cat. Other mountain-bike riders came down this trail, spotted his bike at the side of the trail with no rider nearby. Then noticed a blood trail leading away to a spot where the cat had partially consumed then covered its victim. The bike riders notified police, and this cat was later killed when it returned to the kill for a second meal.

Judy Kimberly, of Tennessee, was recently hunting turkeys in the Sierra Madre mountains and, I suspect, became the target of her own turkey calls. A lion, probably drawn in by the calls, attacked Kimberly in her camouflage hunting attire. Luckily she was armed with a 12 gauge shotgun and ready. She killed the cat at near point-blank range, defending herself.

In Boulder, Colorado, Gail Loveman had a cougar on her back patio in broad daylight. The cat was staring through the glass patio door directly at her terrified pet cat. Gail was able to get a photograph of the cougar making this brazen move. Thankfully her cat was inside, and no children were playing in the yard.

The California Department of Fish and Game reported that there have been a dozen attacks on humans in the state since 1986.

In January of 2010, an eleven-year-old boy in British Columbia was collecting firewood in his yard when attacked by a cougar. This boy, while severely mauled, was rescued when the family golden retriever (appropriately named Angel) came to his defense. Luckily a

RCMP constable was on the scene in minutes to shoot the mountain lion. This cat had to be shot several times to be subdued and still had the golden retrievers head firmly clamped in his jaws even after it died. These cats can be tough.

A cougar attacks with an explosive burst of speed and power. Stray, small 15 pound pet cats can explode when cornered and are impossible to catch or control. To be on the explosive end of a 150 pound cougar is an experience I want to avoid.

A park ranger gave me some good advise from a park ranger regarding cougar encounters (albeit after my one confrontation). The advise was not to run away in a confrontation, avoid any rapid movements, talk firmly to the animal, stand erect, don't take your eyes off the animal, back away from the confrontation, and if attacked, fight back. A passive "play-dead" posture may just get you killed.

Mountain lions can be tough and deserve respect as they are wild animals. In April of 2010, *Reader's Digest* reported that thousands of wild cats are kept as pets in homes throughout the country. Even exotic cats can be purchased for as little as three hundred dollars. The Digest estimated that there are as many as 7,000 tiger and other large cat "pets" in the USA. Mountain lions are wild, period.

My suspicion is that lions, past and present, had a taste for human blood. It is certainly something the Romans exploited in gladiatorial arenas. However cougar attacks on humans appear to be trending upward. The cat has been seen establishing new territory in areas where it vanished decades ago, including very populated areas. Future environmental changes may become more dramatic, and cougars will be fast to adapt/evolve to fit their new surroundings. Change may be underway in North America with mountain lions a component of that evolutionary change.

Chapter Six

Bears are hunters—stealthy, smart, and fearless

(A true apex predator)

North American Bears, black, brown, grizzly, and polar are rightfully ranked at the top of the food chain. They are massive beasts that can subdue large game animals with a single swipe of a paw. When stalking human game, a bear can impose deathly fear with their terrifying hypnotic stare alone.

Black Bears

Black bears are smart, powerful, predators that can run (charge) at over thirty miles an hour. Black bears represent the smallest of the North American bears with adult males averaging 250-300 pounds yet they may be the most dangerous of bears. Black bears have lived in close proximity to humans for such a long time that through association (adaptation) they over came a natural fear of humans. Black bears represent wild, dangerous, and unpredictable animals. In addition to being fast runners with unbelievable acceleration, they can quickly climb trees or swim lakes or rivers. Black bears rely on an acute sense of smell for hunting and protection; it's a sense of smell that is far greater than that of a bloodhound. These bears have good hearing and see with color vision as well as humans. Humans cannot out distance them, out climb them, out swim them, or out hunt them. Compared to black bears, we don't wake up running.

Black bear populations are also on the rise. There are now 750,000 black bears in North America, more bears than anytime in the last 200 years! California's black bear population has grown from 10,000 in the early 1980's to over 38,000 today. New Jersey recently initiated bear hunts to thin the numbers of bears living in Northern New Jersey (very close to New York City). Black bears have moved into areas of New Jersey where they have not been seen in over 100 years.

In September 2011 "Alaska Magazine" reported a study of black bear attacks in North America that revealed most of those attacks were carried out by predatory males not females protecting cubs. An incredible 92 percent of the attacks were black bears actually hunting humans, certainly something to think about.

Brown Bears

Brown bears have a range that covers 200 to 500 square miles, they can weigh over a half ton and stand ten feet tall. Grizzlies are also members of the brown bear family and the polar bear may have even evolved from the brown bear. Brown bears are well equipped hunters. Aside from their massive size, power, and speed, the brown bear has tools or weapons that have evolved to its advantage. A brown bear's teeth can reach three inches in length inside jaws strong enough to crush a human skull with a single bite. Brown bears have claws up to six inches long that are designed to tear things apart. Compared to bears, human weapons are feeble at best (no teeth, no claws, no speed, no power). Only human weapon technology provides us with an edge.

In the wild brown bears are at home and are adapting to life in and near cities. Humans cannot match a bear's ability to detect and avoid enemies with sight, sound, and/or smell. Brown bears can see color much like humans, they hear well, and they have a sense of smell far greater than a bloodhound. They can be cautious, stealthy and unseen when they wish, searching for danger before moving in any direction.

The brown bear population in North America is also on the rise with an estimated 55,000 bears living in the forests, mountains, valleys, and cities.

I was on a moose hunt in Central Alaska for eleven days in 2008. On this hunt I saw several bear signs throughout the woods, scat, tracks, trees clawed, bear tracks, and bear trails but I didn't see a single bear even though I moved through the woods as silent as humanly possible. I wore camouflaged and carbonized hunting gear to conceal my sight and human odor. I did all I could not to be seen, heard, or smelled yet I did not see a single bear. Bears were there but I did not see them because they elected not to be seen. It was their home, not mine. As big as bears are, humans can walk right by them in the woods and bears will remain invisible. They can even hide in open featureless land areas by simply crouching down in a

SAY, YOU'RE RIGHT MOM! THE OLD ONES ARE EASY TO CATCH, BUT THE YOUNG ONES TASTE BETTER.

shallow depression to vanish from sight. Bears are clever, can be very cautious, and often operate a step ahead of humans. Mother bears teach their babies well and if you don't see bears in a forest it does not mean they are gone, bears learn fast how to conceal their presence and stalk their quarry. Never think the crocodiles are gone because the water is calm.

Fortunately, the majority of bears want to avoid human contact even though it gets more difficult as human and bear habitats become one. Much like us bears do have personality and there are good bears and bad bears. An aggressive bad bear is a wild animal that can stalk and kill humans with ease.

Chapter Seven

Bear encounters

(True stories and actual encounters)

The majority of bear encounters in North America are not confrontational. Bears generally avoid humans and humans avoid bears (usually). However with the growing populations of both bears and humans conflicts do happen.

There have been a few humorous bear confrontations such as the bear in Florida that took up residence in a deer hunters tree stand. When the hunter arrived at his tree stand early one morning, he looked up, saw the content black bear sitting comfortably in his stand twenty five feet in the air staring back at him. The deer hunter wisely decided to go back home, maybe mow the grass. In 2010 a bear in New Hampshire broke into a house, ate two pears, a bunch of grapes, took a drink from the family fishbowl (as he tried to fish I am sure) and then departed. It was sort of a Goldilocks in reverse scenario. Another bear in Colorado broke into a parked car to steal a candy bar (they are clever at getting into things and have that great sense of smell). This bear inadvertently knocked the car out of gear sending it rolling down the driveway to crash while the bear, still cramped inside the car, honked the horn. In Estes Park, Colorado a black bear broke into a candy store and was filmed on the store security surveillance video making seven trips into the store gathering candy bars at the check out counter that he took outside to eat. The store owner said, absent the damage to the front door, this bear was very clean and very careful. No mess was left behind however he did eat a lot of candy!

Sadly many bear encounters take a deadly turn.

Deadly Bear Encounters

Author Larry Kaniut has written several books and articles on bears and bear encounters in North America and they are thrilling true-life stories of bear encounters. I have referenced two of his books in this writing as he does an excellent job of presenting the encounters just as they happened, trilling, frightening, and deadly. In his books he presents statistical data on Alaska bear encounters from the early 1900's through 1989. His numbers are qualified only in that they do not represent all Alaska bear encounters. In Alaska's early pioneering history many bear maulings simply went unreported, often occurring in areas where there was no place to report such a tragedy. Brent Kieth, a licensed and registered hunting guide, in Healy, Alaska showed me a photograph he had framed of one of those unreported encounters. In this photograph an early Alaska pioneer was shown sitting on the ground next to what can only be described as a past "grizzly" encounter. The photograph was grizzly in that it involved a grizzly bear and grizzly in that it was deadly. All that remained of a bear/man encounter in this photograph was a skull of a bear next to a skull of a human with a few bones scattered about on the ground. You can only imagine what took place.

In *Alaska Bear Tales*, Larry Kaniut presented a summary of one hundred seventy historical Alaska bear encounters that resulted in forty-five fatalities. Again, this report is only qualified in that it may not be complete with attacks in Alaska's early history going unreported.

The true stories Kaniut presents in his books were given directly to him by the surviving human involved in the bear encounter, witnesses to the encounter when deadly, or verified historical reports. Of these encounters forty-five resulted in fatalities.

Bear encounters in Alaska and elsewhere in North America did not stop after Larry published his book in 1989. In fact, bear encounters with humans seem to be on the rise throughout North America including areas where no past encounters have ever been recorded. Those bear attacks have been reported in newspapers and magazines all the time, here are a few I have taken from local news:

In 2008 a bear turned the tables on bear hunter Aaron Wyckoff, the hunter became the hunted. The bear badly mauled Wyckoff before a fellow hunter saved his life when he shot the bear at point-blank range with a .44 magnum hand gun.

In Moose, Wyoming an Elk hunter was attacked and badly injured by a bear in 2011. Here the hunter may have drawn the bear to his location with elk calls but the attack was so sudden that he had no time to shoot in his own defense. He dropped to the ground and covered his head, living to tell about it.

In Boise, Idaho another elk hunter stumbled onto a resting bear and the bear attacked. He used pepper spray to ward off the bear but did suffer puncture wounds, an injured left hand and a broken forearm.

In 2010 a camper was brazenly attacked in the middle of the night buy a bear and suffered serious injuries. This camper said she initially screamed at the attacking bear but the screaming only made it worse. She instinctively played dead and the bear let loose of his grip.

A backpacker in 2008 shot and killed a grizzly bear that charged him in Denali National Park. Interestingly this attack took place only a short time after people were permitted to carry self-defense weapons in the park.

Near Billings, Montana in 2009 Jerry Ruth, a retired police officer, was attacked by a bear he surprised. This bear was with three cubs, probably felt the cubs were in danger, and would have killed Ruth had he not been able to defend himself with a handgun.

Kevin Kammer was killed and two others injured in Cooke City, Montana by a mother bear with cubs in 2010. The details of this attack are not clear but it is thought that the mother bear was leading her cubs to food (maybe Kammer).

The city of Anchorage, Alaska reported three maulings in 2008 that had residents rightfully concerned. Anchorage is Alaska's largest city with 360,000 people however it also has a black bear population over three 300 and an estimated 50 to 60 brown bears. With bears moving into urban areas, encounters in cities like Anchorage were inevitable.

In 2012 a black bear wandered onto the campus of the University of Colorado and became a celebrity with the students. Wildlife officials however recognized that the celebrity arrangement would lead to trouble for the students and the bear, they tranquilized the bear and moved it to a remote area that had an abundance of oak brush and chokeberry (wild bear food).

In Caliente, California a woman, in a rural area, was out walking her two dogs in 2008 when she was attacked and severely injured by a bear, circumstances surrounding this attack were not clear. Despite severe injuries she did manage to escape and was airlifted to UCLA Medical Center being lucky to survive.

On July 28, 2010 a botanist from Cody, Wyoming was killed by a bear shortly after the bear woke up from being tranquilized by researchers.

A geologist, Robert Miller, survived a grizzly bear mauling near the Iditarod Trail in Alaska. Miller had finished his work for the day and was waiting for a helicopter pick up. Even though Miller was armed with a .357 magnum revolver, he had no time to defend himself as the attack was so sudden. Then after Miller "played dead" the bear returned gnawing and clawing him more. The helicopter pilot dispatched to pick him up had to circle the area several times to drive off the bear before he could land and take Miller to a medical center.

A hiker in Louisville, Kentucky was attacked by a black bear in 2010 that Kentucky State Wildlife Officials claimed was the first recorded bear attack in the state (times are changing). The hiker, Tim Scott, was walking ahead of his wife and son when he saw the bear and was going to call his wife on a cell phone to tell her to take a different trail. Scott said the bear, while not a large one was focused on him, he was lucky to survive.

In April 2006, Susan Cenkus took two of her children with her to view Benton Falls, a popular attraction within Cherokee National Forest when they were attacked by a predatory male bear. The encounter left her and her two year old daughter severely wounded, and sadly her six year old daughter was killed.

Some bear encounters happened tragically when humans did not treat bears with respect as wild animals. People like Charlie Vandergaw in Alaska had fed and befriended over twenty wild bears on his property for years even though Alaska Game Management Officials worked diligently and legally to get him stopped. Game management officials clearly saw this arrangement as a ticking time bomb. As of this writing, Charlie was still hand-feeding the wild bears.

In Columbiana Station, Ohio an animal caretaker, Brent Kandra, age 24, was killed when feeding a "friendly" bear outside its cage in a private zoo. Sadly Brent's father had recently asked him to leave the job.

The most notorious would-be bear naturalist has to be the late Timothy Treadwell. He lived with brown bears each Summer in Katmai National Forest, Alaska filming and interacting with wild bears up close. Treadwell enjoyed celebrity status appearing on late-night TV shows with his videos, taking his "bear message" to local schools, and was featured in film documentaries like Grizzly Man. More responsible naturalists were fearful that Treadwell's inaccurate message to school children was wrong.

Timothy Treadwell thought of himself as a bear's best friend and a savior for these monstrous wild animals. His fool hardy antics with these wild bruins left the outcome predictable, eventually costing him and a friend (who followed his lead) their lives when a predatory boar attacked them in their tent killing then eating both. That predatory brown bear was later found standing over and protecting the devoured human remains when a bush pilot arrived to check on Treadwell.

Another dramatic bear encounter was presented on the National Geographic TV channel. The documentary was entitled Grizzly Face to Face and it was a chilling example of how deadly a large carnivore can be and what they are capable of doing in a split second. In this documentary a trainer was being filmed standing next to a tethered brown bear. The trainer and the owner (Randy and Steve Miller) were cousins and well known by the bear. There were no cubs involved, no feeding issues (this huge bear was well fed), no territory issues, just a photo shooting session of this bear standing next to a person who had been its trainer. The human actions on this day were not new to this bear (no surprise) and there was no reason for this "tame" bear to be aggressive. The trainer walked up to and stood next to this huge upright bear, the trainer was relaxed with his hands in his pockets. The bear just looked down at him then in a flash grabbed his neck in his powerful jaws shook the trainer and he was dead. This attack happened very fast, it was frightening, unexpected, and the man was killed right before your eyes. Bears are wild animals, period.

Change

It was a bear encounter that served as the genesis of this book as bears, to me, represent the ultimate North American Predator. I have witnessed changed bear behavior, observed changed bear habitat, know that bear food choice, and hunting habits have changed. A new relationship with humans may be evolving, ever slowly, as we interact with these bruins each day. These new relationships may simply play a small part on the ever-changing stage of evolution.

Chapter Eight

Evolution Continues

(It is our way of life)

The science behind the theory of evolution is sound. Evolution represents the earth-life process that will continue until the planet is eventually consumed by the sun in a few billion years.

Earth's history includes six major periods of animal extinction over the past 500 million years. With each event the entire living population of earth was decimated, yet life regenerated, never fully disappearing. Evolution's ever-present processes had also proven that the "law of the jungle" did not always prevail. With evolution the strongest may have vanished while those within an animal population able to modify their lives, their behaviors, and eating habits lived on. Those animals able to adapt to our delicate and changing planet were the ones who would then pass on their genes to succeeding generations.

First were the trilobites of the Cambrian period, then the sea creatures, such as acanthostega, moving onto land, adjusting to breathing air and the effects of gravity, in the Silurian and Devonian periods, then to the monster bugs and animals of the Carboniferous Era, on to the dinosaurs of the Jurassic period, to titanis and eventually human evolution. Each animal group has lived in an ever evolving world. Some did well and prospered, some did not and vanished.

Some animal groups vanished due to horrific environmental changes. Dinosaurs were perfectly adapted to and ruled this fragile earth for more than 150 million years. Their domination of planet earth ended 65 million years ago with a big boom, one nasty "slam-bang" and soon the mighty dinosaurs were history. A large comet hit earth in the Gulf of Mexico and the collision changed Earth in such a way that it smothered most life from giant dinosaurs on down the food chain. Most life forms vanished but not all, some life survived, some rebuilt and regenerated. They evolved. Interestingly insects were among the few survivors.

Talking monkeys

"If humans evolved from monkeys and apes,
why do we still have monkeys and apes?"

Comedian George Carlin

Humans do fit into this evolutionary melee and we have our own body parts that serve as reminders. Humans are born with a tail bone and that is exactly what it is; a "tailbone." Our body once supported a tail and it might have been a very useful tool. The tail may have served as a limb to hold things, a device for cover, or a tool for balance and agility, as with the cougar. Regardless, just as forearms ceased to serve the T-Rex, our tails became a tool that, over time, became useless and has all but vanished.

Humans also have ear muscles that can make our ears move. We do not use those muscles as they don't do anything for us, absent a few clowns employing them for entertainment by wiggling their ears. Those muscles however may have had some human historical importance and continue to serve other animals very well.

Think about a fox, a deer, or a coyote. Those animals use ear muscles constantly that have proven to be a very important adaptation for survival (for hunter and hunted). A wild animal can move its ears to better hear and locate a noise. This helps them hear and identify a predator or prey by only moving their ears. They do not give away their position as they might by moving their entire head or body. In the wilderness movement will quickly give away any animals' location. By only moving their ears an animal can operate like a radar station taking in information needed to avoid predators or to launch an attack on prey without detection. Hearing, locating, and tracking noise has been a very important evolutionary animal adaptation.

A philosophy professor my son, Eric, had in college was fond of referring to humans sarcastically as just a bunch of "talking monkeys" warning that talking monkeys may face a limited future on the planet. This professor advised that talking monkeys were clever but too engaged in inventing ways to mass-murder other talking monkeys. The professor was not an optimistic individual. I do remember Eric talking about this professor's views on shared human anatomy (tail bones with apes) including an infant's strong hand grip ability. A human baby is born with a very strong grip for the first few months following birth. Hand muscles working effectively before any other body muscles develop. Behavioral scientists believe this too is a carry over from the distant past when a strong grip was needed by new born babies to hold onto a mother for protection and transportation. The talking monkey professor made some valid points.

Trillions and trillions of animals have lived on earth, calling it home and trillions still do. Which animals will prevail going forward is that mystery called evolution and it takes place all around us all the time. It may speed up or slow down but it never stops.

Think about this

What if "reverse evolution" took place and we ended up living in some type of "Planet of the apes" scenario (my son's philosophy professor might say we are already there)? Certainly not likely however massive animal life changes do happen on earth as six major periods of animal extinction have taken place and many more minor periods of extinction. With each cataclysmic event, life forms regenerate (evolve) changing and adapting. Considering earth's dramatic ever-evolving history, anything can happen. It's something to consider.

Earth once had a single land mass called Panaea that came together around 300 million years ago during the Mississippian period and the start of the Carboniferous Era. Coming together means that continents were not joined, became one, and then about 200 million years ago, during the Jurrasic period and the end of the Carboniferous Era, started to drift apart. Scientists also advise that the continental structure we have now will eventually return to a single land mass. All this continental drift is like a giant slow-motion version of the dodge-'em cars at an amusement park. All together in one mass then slam bang against each other as the ride starts out in a controlled sphere, then slam-bang all stopped and together again. Earth's unending continental drift will create land-slides, tsunamis, earth quakes, volcanic eruptions, and other events that will alter the environment and potentially change life on earth. An ape planet may be unlikely however change is inevitable.

Truth is, earth is a very delicate planet resting within a massive universe and any number of catastrophic events can take place at any time to throw the life-cycle equilibrium out of balance.

Life evolved on earth, animals lived and animals died, entire groups of animals have vanished yet animal life on this remarkable planet continues. Eventually, in a few billion years, all life will be gone when Earth's magnificent experiment is ended.

Epilogue

The Universe and Beyond

What does the future hold?

Viewing stars in the night sky has fascinated humans since the beginning of time, and this writer shares that fascination. Out there are billions and trillions of stars with over 300 billion in our Milky Way Galaxy alone. Most of these stars harbor their own solar system some very similar to the one earth resides in. These stars, in turn, reside within galaxies, and those galaxies reside in our universe, a cosmos, that exceeds clear scientific definition or human imagination.

"Just remember this my friend,
when you look up in the sky,
you can see the stars, but
still not see the light"

Song: "Already Gone"
The Eagles

A light-year is defined as the distance light travels in a year. Light speeds along at 186,000 miles per second, and a year has three hundred sixty-five and one quarter days, consequently the distance of a single a light-year is 5,578,625,000 miles (5.6 trillion!). One single light-year represents a distance that exceeds my ability to fully comprehend, and our universe is trillions of zillions of these light years in diameter and growing (one Big Mama). Science cannot honestly define the distance, cannot say with certainty that there is another earth-type planet, nor can science even tell us if ours is the only universe. There just might be a sister universe, and if some theorists are right, there may be millions of universes out there. Do those universes operate on the same rules of physics or science? Is atomic structure going to be the same with other universes? So very big, so many questions, so few answers, so much to learn. With our limited human knowledge, the only certainty is that what we don't know is exponentially greater than what we do know. I suspect it will always be that way.

In our tiny little part of this great universe our solar system consists of a sun and nine planets: Mercury, Venus, Earth, Mars, Jupiter, Saturn, Uranus, Neptune, and Pluto (I still count Pluto). Four of these planets are bigger than earth with the giant Jupiter at eleven times the diameter of earth. However, if you combined all these solar planets with the sun into one big monster, the sun segment of that monster would represent 99.85 percent of that mass. Damn, it's big, and our sun is only a medium size/medium age star. It would take a million earths to equal our sun and still there are stars in our galaxy that are a million times bigger than our sun. Our entire solar system is a meager speck of light in space, sitting within the massive Milky Way Galaxy, one hundred million light-years wide. While the Milky Way Galaxy contains billions of stars and billions of solar systems, even this massive galaxy represents a mere dot in cosmic space . . . a miniature dot.

Humans have historically considered our earth and area in space as huge beyond imagination, yet it is a microscopic speck of dust in the cosmos, and the entire time taken for human evolution to unfold on earth is a fraction of a split second in solar time.

Science and Religion:

In human eyes, life on earth is a marvelous blend of science and religion. Science cannot answer every question, however it continues to acquire more knowledge and opens new doors of learning. Conversely, many people modify, or re-explain, their system of belief as new scientific discoveries are made. Debates are endless regarding religion, and I don't care to get into any religious debate, however I will submit that most humans have some system of belief. I say this because religion is a philosophy to live by, and in basic terms, philosophy is a person's belief or view of life and death. All humans have some view/ belief regarding life and death, consequently humans are philosophers and being philosophers (crude as my analogy may be) have some religion . . . at least in very basic terms.

It is here, religion, where human animals (talking monkeys) might just differ from all other animals. Wild animals can think and reason much the same as humans. Bears can analyze a situation before taking action, they learn to understand if "A" and "B" occur in a situation that "C" may result. Here too the more senior, mature, and experienced bears are so good with reasoning ability they can often out-smart humans in many situations. The big behavioral difference is that humans have evolved to understand that one day they will die. Unlike wild animals, we try to bury our dead, remember them, and have opinions on how life and death transpire. We are philosophers and in my analogy, animals with religion, something bears don't have.

While science cannot have answers to all our many questions, through scientific research, we have been given a greater understanding of our universe. It is clear that the universe was created by "something" that is great and powerful. That is where religion comes into the picture. There is a power existing in our universe(s) that is far greater than man and beyond what science can explain. That power cannot be ignored. Call it your God, call it a power, call it the force, call it whatever you wish, but it is undeniably present, it is colossal, and it controls everything, including the continued existence of the entire planet earth.

Who's in charge?

With all the wondrous, challenging, and disastrous events taking place on earth it is fair to ask a few powerful questions:

Who's in charge of the ant hill?

Who's in charge of the bears?

Who's in charge of the earth?

Who's in charge of the universe(s)

Talking monkeys are clever and have evolved to ask questions and seek answers, ask more questions and seek more answers. This process never ends.

Future:

What does the future hold? Damn, I wish I knew. For the near term or long term, you need to make your own decision. *Who's In Charge* only takes a glance at some minor present day wilderness changes I have detected that, microscopic as they may be, are potentially part of a much bigger picture, you decide.

Each day is a beautiful gift from our God, the power, the force, whatever that giant "something" is out there that controls the universe(s) we live in and built this dynamic, ever evolving, planet. Again, that's your call. One day, hopefully well into the future, everything we have on our delicate planet will certainly be taken away, in the distant future perhaps by known or unknown cosmic forces and in the near, foreseeable, future maybe, just maybe, by those pesky bugs and insects.

For the near term I keep thinking: Hey, Bug Boy might have been right.

References

Alaska Bear Tales, Larry Kaniut, Alaska Northwest Books, 1983

"Coyotes are the new top dogs" by: Sharon Leng, Nature Magazine.

Earth Science, 11th Edition, Edward J. Tarbuck & Frederick K. Lutgens, Prentice Hall,

Upper Saddle River, NJ

Encyclopedia of Insects and Spiders, edited by Christopher O'Toole, Firefly Books, Buffalo, NY 2002

"Evolution in coyotes (canis latrans) in response to the megafaunal extinctions", Julie A. Meachen & Joshua X. Sameles, National Evoluctionary Systhsis Center, Durham, NC.

A Field Guide to the Insects, Donald J. Barron/Richard E. White, Houghton Mifflin Co, Boston/NY 1970

Funk & Wagnals New Encyclopedia, Editors: Leonl. Bram, Norma H. Dickey, and Robert S Phillips, Funk & Wagnals Inc, NY

More Alaska Bear Tales, Larry Kaniut, Alaska Northwest Books, 1989

National Geographic Documentary (aired 2011-2013), *Grizzly Face to Face.*

Never Cry Wolfe, Farley Mowat, Little, Brown, & Co., Boston—Toronto, 1963

Webster's New World Dictionary, Fourth edition, Michael Agnes, Editor in Chief,

Pocket Books, NY, NY

Wikipedia, The free encyclopedia, On-line.

Wolves at our door, Jim and Jamie Dutcher with James Manfall, Pocket Books, NY, NY, 2002.

World Animal Foundation Coyote fact sheet. On-line.

Reader comments are welcomed.

Please send to:

Bill Wonders

President

Woods and Water LLC

PO Box 381

Burton, Ohio

44021

Other books by Bill Wonders:

Orion Too

Is a business/people management book with a unique approach. Author shares with readers an adventurous outdoor hunting journey into the "Last Frontier" of Alaska. He takes time throughout the journey to reflect back on over 35 years of human resource management. Reflections include comments on the many positive changes that have taken place in business.

The book is written in a humorous and conversational way readers will find enjoyable. Many "real-life" work examples are provided that support principles of disciplined, cost effective and professional human resource management.

In business we are all people managers and too often we miss opportunities to extract the best work performance. Often these opportunities cost little financially yet secure outstanding performance results. Author submits these people management ideas with actual work site examples.

This book is fun to read and yet provides a great learning opportunity.

Machine Shop Poet

This book is a presentation of poetry written by William W. Wonders. Much of his poetry has been lost over the years. He would write poetic verse to commemorate an event, recognize an accomplishment, or tell a friend or loved one just how much he cared. Once written and delivered, only a few copies were retained, Machine Shop Poet is a selection of those poems. Several poems presented were only found after his passing;

beautiful verses that no one had ever seen. These poems come at you from the heart and reflect values of family, friendship and love.

About The Author

Bill is a life-long outdoors man and experienced naturalist. His wilderness explorations have covered North America from coast to coast and from Florida to Alaska. He holds a bachelors degree from Kent State University and a masters degree from Youngstown State University. He currently serves as president of Woods and Water LLC with headquarters in Burton, Ohio